ULTIMATE BOOK GUIDE ON MOBILE REPAIRS AND INSTALLATIONS

I0454381

Beginners Step-by-step for repairing phones, systems and desktop

Arthur K. Henry

Table of Contents

CHAPTER ONE

INTRODUCTION ON MOBILE REPAIRS

Most persons find it difficult in repairing mobile devices due to lack of solid in-depth knowledge of how to repair devices; as a beginner or owners such devices you can guide yourself with help guide of this book.

INSTRUMENTS TO REPAIRS

A FASTENING IRON

A fastening iron is utilized to patch little parts like capacitor, resistor, diode, semiconductor, controller, speaker and mouthpiece, show and so forth.

A PATCHING STATION

It is a multipurpose power obtaining fastening gadget used to bind various parts. A patching station is an essential piece of the cell phone fix tool stash since it furnishes with required welding office when required. A patching station comprises of two units: station and iron. It has the choice to manage temperature in view of the binding intensity necessity. The patching iron gets gotten together with the welding station.

A BIND WIRE

Bind wire is utilized to weld electronic parts, ICs or jumper.

A MULTIMETER

A multimeter can be simple or computerized. In cell phone fixing, generally a computerized multimeter is utilized to find flaws, really look at track and parts.

AN ANTISTATIC MAT (ESD MAT)

An ESD Mat or Antistatic Mat is laid or put on the table or workbench where portable fixing is finished. The mat is grounded utilizing an establishing string or typical establishing wire. This keeps harm from friction based electricity.

A MAGNIFYING INSTRUMENT OR MAGNIFIER

These used to see an amplified perspective on PCB or electronic parts.

AN AMPLIFYING LIGHT

Seeing the amplified perspective on the PCB of a cell phone is utilized. Most amplifying lights additionally have light.

HOT AIR BLOWER

A hot air blower is likewise called SMD (Surface Mount Gadget) revamp station and SMD fix framework. It has control to direct or oversee temperature and stream of hot air.

ACCURACY SCREWDRIVER

Accuracy screwdriver is utilized to unscrew and eliminate and fix screws while collecting and camouflaging a cell phone.

SCREWDRIVERS

A screwdriver is a piece of each tool compartment. It is a device with a smoothed or cross-molded tip that squeezes into the head of screws. While in cell phone fixing, screwdrivers become accustomed to fix or relax minor-sized screws that are difficult to hold with traditional measured screwdrivers.

THE CELL PHONE OPENER

These are utilized to open the lodging or body of a cell phone.

THE ESD-SAFE CLEANING BRUSH

These are utilized for cleaning the PCB of a cell phone while fixing.

PCB Holder or PCB Stand

A PCB PRINTED CIRCUIT BOARD

Holder or PCB stand is utilized to hold various sorts of PCB of a cell phone while binding or fixing. It holds the PCB unequivocally and doesn't permit it to move accordingly helping in fixing.

THE TWEEZERS

Tweezers are leveled molded gadgets contained two jaws used to get little items like jumper wire, coordinated circuits, and so on, in patching or de-binding. It is a high priority device for the cell phone fix pack. Fundamentally, while fixing the central processor of the versatile, tweezers are utilized.

THE DC POWER SUPPLY

Controlled DC Direct Current power supply is utilized to supply DC current to a cell phone. Most fix individual utilized DC power supply to turn ON a cell phone without battery. It can likewise be utilized as a battery sponsor to help battery of a cell. It can likewise be utilized as a multimeter.

THE BATTERY ANALYZER

This gadget is utilized to test and dissect status or state of battery of a portable phone.

THE TEST DANCE BOX

This gadget is utilized to analyze and track down shortcoming or issue in a cell phone. It assists the cell phone with working and capability regularly outside its case or lodging.

THE LCD ANALYZER

LCD Analyzer is utilized to check regardless of whether LCD screen of a cell phone is defective.

CHAPTER TWO

WORKING DEVICES GUIDANCE'S

WHEN CELL PHONE BATTERY IS NOT CHARGING AND SOLUTIONS

Change the charger and check. Voltage should be 5 to 7 Volts. Clean, Exchanged or Change the Charger Connector. If the issue isn't addressed then change the Battery and Check. Non Removable Battery is Simply Stuck at the Base. They can be Eliminated Effectively however little cautiously.

Check Voltage of the Battery Connector utilizing a Multimeter. Voltage should be 1.5 to 3.8 DC Volts. If there is no voltage in the connector then check track of the charging segment, Allude to the chart of the specific model of the cell phone.

If the issue still not fixed then actually take a look at Circuit, Curl and Controller individually. Change whenever required. If the issue is as yet not tackled then Intensity or Change the Charging IC. Lastly Intensity, Reball or Change the Power IC.

GUIDANCE IN FIXING NO SOUND FROM REMOTE EARPHONE

Press Power + Volume Down

Switch On the Earphone; Simultaneously Press Power + Volume Down Buttons for 7 to 15 Seconds.

You can likewise Press the Power Button for around 7 Seconds in the territory of Closure to Reset. Now the Earphone is reset. Manually Pair the Earphone with your Cell phone interestingly. From Second Time, they will get Matched Consequently.

Reset Utilizing Committed Key Button

A few earphones have committed Key Button for Processing plant Reset. Simply Press the Key for 7 to 15 Seconds to Reset and Fix the Issue.

Reset Utilizing Sim Ejector Pin

A few earphones have devoted Little Opening for Processing plant Reset. Simply Press involving Sim Ejector Pin for 8 to 15 Seconds to Reset and Fix the Issue.

IOS 7 NOT ANSWERING

If the iPhone, iPad or iPod isn't answering then attempt to restart it; to restart the gadget, Press and hold the Rest Wake button on the top until the red slider shows up. Presently drag the slider to switch the gadget OFF. Presently press and hold the Rest Wake button again until you see the logo.

Turning

If the gadget doesn't answer by any means or doesn't turn on, then, at that point, attempt to reset it. To reset the gadget, press and hold both the Rest Wake and Home buttons for around 12 seconds. At the point when the logo shows up, the gadget is reset. If any single application isn't answering or quits answering then attempt to drive stop the application. To drive close an application, double tap the Home button and Swipe left or right to find the specific application.

Point

Presently swipe the application's see up to close it. If your concern neither is nor settled by any of the above arrangement, then, at that point, you might attempt to reestablish your gadget with iTunes.

ISSUES RELATING TO SCREWDRIVERS INAPPROPRIATELY

ISSUE ON HAND INJURY

Utilizing some unacceptable method or applying extreme power while taking care of a repairman screwdriver can prompt hand wounds like strains, rankles, injuries, and, surprisingly, more serious issues like breaks. The redundant movement engaged with utilizing such a device can cause a ton of weight on the hand muscles and joints.

ISSUE ON EYE INJURY

While working with a repairman screwdriver, there is a gamble of metal shavings, residue, or flotsam and jetsam taking off suddenly. Without appropriate eye security, these particles can cause eye bothering or even serious eye wounds in the most pessimistic scenarios.

ISSUE ON ELECTRIC SHOCK

While dealing with electrical gadgets or machines with a screwdriver, there is a likely risk of electric shock in the event that the client comes into contact with uncovered wiring or live circuits because of ill-advised use. This can be possibly lethal if the client don't watch out and doesn't go to satisfactory wellbeing lengths.

HOW TO HANDLE SCREWDRIVER APPROPRIATELY

USE THE RIGHT SCREWDRIVER DIGIT

Continuously ensure that you pick the right screwdriver digit that matches the screw head. Utilizing some unacceptable sort can bring about stripped screws or harm to the screw head.

PROPER HOLD THE SCREWDRIVER ACCURATELY

Hold the screwdriver immovably and easily with your hand. Try not to utilize unreasonable power, as it can prompt slippage and likely wounds. It is likewise critical to keep a steady stance. If vital, utilize your other hand to help the item you are chipping away at.

TIE WEAR WELLBEING GOGGLES

It'd be smarter to wear wellbeing goggles to safeguard your eyes while working with a technician screwdriver. This will safeguard them from trash and decrease the gamble of eye wounds.

WHAT TO KNOW ABOUT HALF SHORTING AND FULL SHORTING

Half Short or Half Shorting

Half Short or Half Shorting of a Cell Phone PCB is the Condition when a Telephone Gets Switch ON and it likewise works however the Battery Channel Extremely Quick. This truly intends that there is some dry patch on the PCB of the Telephone or some little Part, for example, a SMD Capacitor is Flawed or Short which should be taken out or Supplanted.

WAYS TO AVOID HALF SHORTING

Keep the Advanced Multimeter in Ringer Mode. Check the Battery Connector of the Telephone in Forward Red Test on Sure and Dark Test on Negative. On the off chance that the Worth on the Multimeter is 1, the PCB is alright and there is NO Shorting.

Now really take a look at the Battery Connector in Switch Red Test on Negative and Dark Test on Certain. Assuming the Worth on the Multimeter is Between 250 to 600 there is NO Shorting and the PCB is alright. If Worth shows up at Both Forward and Invert Checking, then, at that point, there is Half Short.

Full Short or Full Shorting

Full Short or Full Shorting of a Cell Phone PCB is the Condition when a Telephone doesn't get Switch ON and the Telephone is dead. This implies that at least one than 1 of a few Significant Electronic Parts is flawed and the Board should be checked completely to fix the Issue.

WAYS TO AVOID FULL SHORTING

If there is Signal Sound in Both Forward and Switch predisposition then there is a Full Short._If we interface Positive (+) and Negative (-) of the Battery Connector to a DC Power Supply Machine, Ampere (A) would Begins to Diminish without Turning ON the Telephone, Then, at that point, there is Full Shorting in the PCB of the Telephone.

SOLUTION TO FIX SHORTING IN CELL PHONE

To Fix Half Shorting, The Initial Step is to dismantle the Telephone and Clean the PCB Completely with IPA Arrangement and the

Apply Intensity All around the Board utilizing Hot Air Blower Machine. Get together Back the Telephone and Check in the event that the Issue is Settled or Not.

Generally speaking, the Issue gets settled. Yet, in the event that there is still Half Shorting, Check for Little SMD Capacitors Close to the Battery Connector. Eliminate the Broken One. On account of Full Shorting, Interface the Positive (+) and Negative (-) of the Battery Connector to a DC Power Supply Machine. Amp will Begin to Diminish without Turning ON the Telephone.

POWER SUPPLY

This implies there is Full Shorting in the PCB of the Telephone. For this situation, interface positive and negative of DC power supply to the battery connector, then, at that point, diminish the voltage to nothing and afterward increment the voltage gradually to the limit where Ampere begins to diminish.

Keep by then for a couple of moments and check in the event that any part is getting warmed. You can experience the Intensity of the Flawed Part with your Fingers. In the event that any part gets warmed, it should be checked and supplanted.

BASIC GUIDE IN FIXING MOBILE PHONE FAULTY

Open the menu and select settings. You will track down a possibility for reset settings in this menu. Select this. In this you will get different choices viz. Reset settings just and reset all ace reset; eradicate all information. Select the subsequent one. You will be requested the security code, In the event that you have changed the default code, type the changed code while possibly not then enter the default security code . On the off chance that you don't have the foggiest idea about the default code then click on default security codes for all telephones and get the code for your kind of versatile from that point and type in. Your telephone will be designed.

BASIC GUIDE ON SYSTEM UPDATE (FIX APPLICATION IN WINDOWS)

Open the Settings application. Press the Windows + I console alternate route. Go to Refresh and Security. Select the Recuperation tab. Under High level Startup, click Restart Now. After the PC reboots, go to Investigate > High level Choices > Startup Fix. After the startup fix process is finished, reboot the PC. If the PC neglects to boot into the working framework, start startup fix utilizing the Windows establishment circle.

THE BASIC GUIDE ON LOCK SCREEN

Click Beginning. Click the power button symbol. Press and hold down the Shift key and afterward click Restart. Keep on holding the Shift key until the PC reboots. Select Investigate. Go to Cutting edge Choices > Startup Fix. After the startup fix process is finished, reboot the PC. If the PC neglects to boot into the working framework, start startup fix utilizing the Windows establishment circle.

ANALYSIS ON INWARD LINKS AND CONNECTORS

The links utilized in computers end in various connectors. By show, each connector is viewed as one or the other male or female. Numerous male connectors, additionally called fittings or headers, have projecting pins, every one of which guides to a singular wire in the link.

The relating female connector, likewise called a jack, has openings that match the pins on the mating male connector. Matching male and female connectors are joined to frame the association. A few links utilize unsheathed wires joined to a connector. Three links of this sort are normal in computers those used to supply capacity to the motherboard and drives; those that interface front-board LEDs, switches, and at times USB, FireWire, and sound ports to the motherboard; and those that interface sound out on an optical drive to a sound card or motherboard sound connector.

www.ingramcontent.com/pod-product-compliance
Lightning Source LLC
Chambersburg PA
CBHW060019300526
45794CB00003B/1218